DATE

DATE

DATE

DATE

DATE

DATE

DATE

DATE

DATE

DATE

DATE

DATE

DATE

DATE:

DATE

DATE:

DATE

DATE:

DATE:

DATE:

DATE:

DATE:

DATE

DATE

DATE

DATE

DATE:

DATE

DATE

DATE

DATE

DATE:

DATE

DATE:

DATE:

DATE:

DATE

DATE

DATE:

DATE:

DATE

DATE:

DATE

DATE

DATE:

DATE

DATE

DATE

DATE: ______________________

DATE

DATE

DATE:

DATE:

DATE

DATE

DATE:

DATE

DATE

DATE

DATE:

DATE:

DATE

DATE

DATE

DATE

DATE

DATE

DATE

DATE:

DATE

DATE

DATE: ....

DATE

DATE

DATE:

DATE

DATE

DATE

DATE

DATE

DATE

DATE

DATE: ____________

DATE

THE 99 DAY CHALLENGE - YOUR WAY TO SUCCESS!

Losing weight has never been easy and is almost always associated with deprivation. At the same time, however, it is also the way to more well-being, a better body feeling or better health. Only those who persevere will benefit from the positive effects, but all too often the motivation is lost and we give up without receiving the reward for our hard work.

So that this does not happen to you, we have elaborated this weight loss journal. When it comes to losing weight, we often see only the long, strenuous path that lies ahead and thus create the hurdle that prevents us from starting with the project. Here you have the opportunity to record all the achievements, even the smallest ones, and always keep your progress clear. There are almost daily small rewards on the way to our dream weight, which compensate us for the effort and continue to motivate us. Not only achieving the goal makes you happy, but also these many, small victories waiting for us at the edge of the road.

SUPPORT YOURSELF

Children grow through love, that is nothing new. Through recognition and appreciation they develop into good and happy adults. If, on the other hand, you were to yell at them all the time, just telling them what they did wrong while ignoring what they did right and tell them every day how ugly they are, their magical spark will disappear very soon. They would have no motivation to do something good and correct because they would not get any appreciation and recognition for it anyway.

Now, imagine how you speak to yourself. Do you look into the mirror and think “You‘re so fat, you become uglier every day“ or “You‘re so weak and poor, have you been eating the whole bar of chocolate again”? Never forget:

“WHETHER YOU FIND YOURSELF FAT OR EVEN OBESE, IT'S NOT WHAT YOU ARE. FAT MAY BE SOMETHING YOU HAVE, BUT IT IS ONLY A PART OF YOU AND BY NO MEANS WHAT DEFINES YOU AS A HUMAN BEING!”

Praise and recognition are the seeds of motivation to change, even in adults, remember. So start to value yourself today!

WRITE DOWN YOUR 10 MOST LOVABLE FEATURES HERE.

•
•
•
•
•
•
•
•
•
•